AuthorHouse™ UK Ltd.
500 Avebury Boulevard
Central Milton Keynes, MK9 2BE
www.authorhouse.co.uk
Phone: 08001974150

First published by AuthorHouse   2/24/2011

ISBN: 978-1-4567-7532-2 (sc)

authorHOUSE®

This book is dedicated to:

my mother **Purabi Moulik**

– Subhendu Moulik

_____

my wife **Marianne**

– Frank Feely

Acknowledgements

We are grateful to
Mr Sinu Mathew Zachariah
and
Mr. Sunil Varghese
for their suggestions and support
in composing this book.

Preface

A little about the Authors, we are from different backgrounds, one of us from Europe the other from Asia, one from a Project Controls discipline and the other from Engineering. This book is the result of a collaboration that began with working together in Japan to resolve contract changes on a major oil and gas project.

Neither of us is a professional estimator but both of us are involved in estimating day to day. It was evident to both of us that an appreciation of what goes into producing an estimate was missing from a lot of engineering and construction professionals "education". This often results in a resistance from these disciplines to get involved in what is seen as "bean counting" or a complete overreaction when presented with the total cost of the latest change in design they require. Our experience is mainly in the oil and gas business and we are using that experience here so references will be related to that. That said, we believe that estimating principles are applicable to any industry.

Therefore we conceived the idea of writing a book that is meant as a guide to non-professional estimators to give an insight and an understanding of the complexities of estimating total project costs.

Estimating is not a new discipline and there are many books available written by estimators for estimators. They will provide nuts and bolts knowledge of every aspect of estimating. There are many Consultants who specialise in servicing the need for ever more accurate estimates. The purpose of this book is not reinventing the wheel or trying to outdo these sages and tomes but to give an overview and some basic guidelines in a very simple way.

Introduction

This book will give you the basis of what makes up the costs of building a plant and how they are estimated.

There are many factors that affect these costs and they are so varied that it is impossible for a simple book to cover every issue, but hopefully there is enough information within these pages to identify the key issues that have to be addressed and the factors affecting them.

Why estimate the costs? Why not just decide what you want to build and get on with it? After all the costs will be what they will be, surely? This is often a common attitude held by the uninitiated. Things are not so simple though. In everyday life we use cost estimation as one of the factors in making decisions. Whether it is deciding to buy a magazine or using the money to buy a sandwich or wondering if we should buy a new car or a used one it is important to have an idea of what the costs are likely to be and get a feel for what sort of value for money we will get for our expenditure. The same is true for any project undertaken in any industry. No one will invest money unless they have an idea of the costs involved versus the profits to be made or benefits to be gained.

To this end companies and organisations put a lot of emphasis on and spend a great deal of time, energy and money in getting the most accurate estimate they can for what they want to do.

Contents

Appendices

Chapter **1**

What is **Cost Estimation?**

Cost estimation is the process by which the cost of doing something is assessed before it is completed.

So, are estimates just good guesses? The answer is no. There are a myriad of methods, mathematical models, scientific principles and engineering techniques which can be applied to the simplest level of cost detail to give the best chance of arriving at as accurate an estimate as possible. But the very word estimate implies that there will be a level of uncertainty involved.

Besides calculating the costs of the actual work to be done the uncertainty has also to be estimated and allowed for.

There is a difference between a cost estimate and a cost forecast. When talking of project costs these often blur together, but generally a cost estimate is what you have before you commence building your project and a cost forecast is what you have once you begin constructing your project. Or, a cost estimate is what you use to decide to commence the project and a cost forecast is what you use to control the project during execution.

For the purposes of this book we are looking at what is required to estimate the total cost of a project during its different stages from the concept to the start of construction. That is not to say of course that estimation ceases once a project begins, not in the least. Throughout the course of the project there will be the need to provide estimates for various purposes, not least of which will be scope additions or deletions, process changes or correcting design errors. In this

regard the role of engineers and project managers becomes vital as often they are the estimators. It is therefore vital that they consider all elements of the costs that will be incurred. Often we have seen estimates given by engineers which bear no resemblance to the final actual cost and usually the blame for this is passed to the commercial disciplines "they paid too much", when in fact it is the failure to include all cost elements that is usually the problem. Technical specialists can be blind to the indirect costs that attach to the direct scope they are looking at.

Naturally projects need to be planned and managed within scope, time, and cost. These elements are interrelated. Change one and you affect the others. For example, increase the scope and you will increase the cost and time. Perhaps time is the major constraint in the execution of the project so therefore the need to complete the project quickly will result in working overtime or nightshifts which will increase costs. A further issue to be considered in developing any cost estimate is the execution plan. That is the way the

ESTIMATE

project will be achieved. In particular the contracting strategy will affect the estimate. There needs to be specific attention paid to allowing for the risks involved with the different ways that a type of contract can generate costs not directly related to the volume of work to be performed. This subject will be dealt with in the chapter on Risk.

In order to assist project owners and managers to make decisions on whether to proceed with projects accurate estimates are required. From these estimates, rules of thumb can also be generated to help evaluate future estimates or allow project managers, engineers, to check the estimates they review or need to generate. The method of producing these and a worked example can be found in Chapters 7 and 8.

Chapter **2**

Types of **Cost Estimation?**

The reason that an estimate is required is usually because someone somewhere has had a bright idea to make money.

During the life of the project various cost estimates are carried out depending on the stage at which the project is at. Obviously the more information that is available the more accurate and detailed these estimates will be. These estimates will typically be given an accuracy range and this is often referred to as the "class of estimate" an example is a 50/50 estimate. This implies that the costs could be 50% higher or lower than the estimate and is very preliminary. As the project progresses these ranges will reduce until at the end of detail engineering, after completion of flexibility analysis, final material takeoff, a detailed cost estimation is carried out and an estimate accuracy of +/-5% should be achieved. When these estimates are carried out and exactly what accuracy levels are applied and how the various estimate classes are named varies from organisation to organisation, but the principles are the same.

Generally in projects, three types of cost estimations are carried out. If you scan through all the available literature and information on the Internet you will read that there are 2,3,4, or even more types of estimate but we reckon through our experience that three is the number.

Preliminary, scoping or screening estimate
Detailed
Final

Preliminary or screening estimate

The reason that an estimate is required is usually because someone somewhere has had a bright idea to make money. Whether that is by building a new plant to increase production capacity or altering a plant to improve production or whatever, somehow we need to know if it is worth considering.

This is the stage at which screening or feasibility studies, which as the name implies determine if the project is indeed feasible, take place. For these we need to have an estimated project cost which can be given rapidly and will not cost a lot to produce. At this time there is no project design data to rely on. This cost estimate is therefore usually linked to the production capacity of the plant (and often referred to as an order of magnitude cost estimate) and will be based on historic final costs of similar projects scaled up or down relative to the expected production capacity of our plant compared to the capacities of those similar projects. To make this estimate more accurate we will have to consider the expected location of our plant which will have an influence on land and labour costs for instance. We also need to know when the plants we are using for reference were completed so that these costs can be escalated to current values. The sequence in which we develop this kind of estimate can greatly affect the accuracy. If the

location for our plant has not been decided then we need to establish a base cost and work up some factors to apply to the bottom line for the various areas of the world we may want to build our plant. This will include a consideration of the area/country's economic and political stability and for these issues we will have to apply certain estimated amounts which we will put in the project risk account. Risk and Contingency will be discussed in a later Chapter. The accuracy of this type of estimate is typically no more than +/- 35 to 40%. There are many issues which will be considered in deciding to go ahead with the project, including sales data etc.

Detailed

Before taking a final investment decision, the plant owner will seek a detailed cost estimate from the estimator. During this time basic project design information will be available e.g. Plot plan, Flow Diagrams, P&IDs, major equipment is sized and metallurgy of pipe and equipment is specified, pipe-rack information, major pipe routings , plant power requirement , number of substations, generators, Control buildings and extent of instrumentation. The estimator will use the equipment cost and main bulk components to generate the estimate. These cost elements are often referred to as prime costs. Often these prime costs will be factored to produce a more detailed estimate than the screening estimate. This estimation has an accuracy of +/- 20-25%.

Final

Towards end of engineering, another detailed cost estimation exercise is usually carried out once again. This is in order to determine a more refined estimate. This estimate will be used to control the costs during execution and is often referred to as the control estimate. During this estimate, the bulk material take off gets firmed up and any change in metallurgy, addition/deletion of tagged items gets captured. Execution plans and contract strategies are known and fixed and bids are in for all the major elements of work. From here on in we are in the Forecasting phase of the project.

Estimate Accuracy

It may seem entirely obvious but the more information that is available to the estimator the more accurate the estimate will be.

This accuracy needs to be quantified because it is one of the components of the estimate, it is the metric that will determine part of the contingency. Below is a graphical representation of the way the accuracy is affected by the stage at which the estimate is performed. Most organisations have a corporate standard version of this or similar which is used by their estimators to develop the contingency.

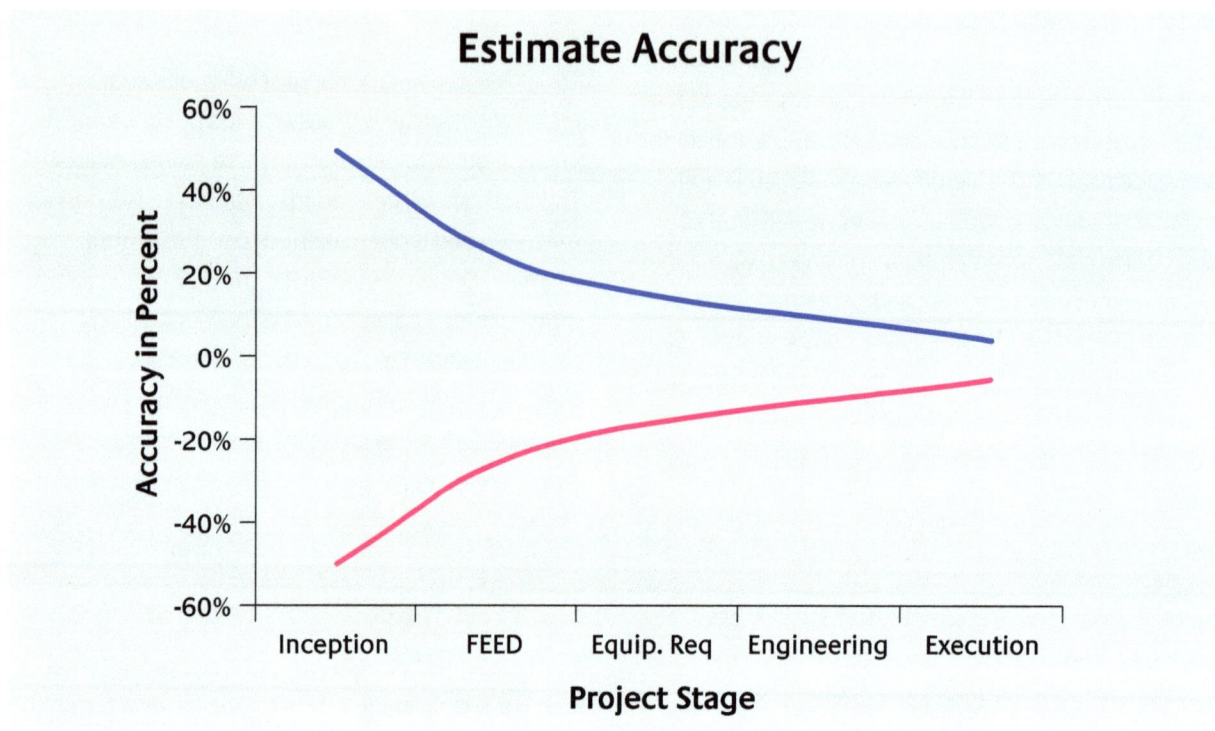

When these estimates are carried out and exactly what accuracy levels are applied and how the various estimate classes are named varies from organisation to organisation, but the principles are the same.

Chapter 3

Components
of Cost Estimation

In this chapter we will look at the components of a cost estimate and describe what they are. Basically costs are divided in to two main categories, prime costs and derived costs. Prime costs are those for which a cost can be directly calculated (because we can phone suppliers for unit rates for a given specified item but we may not know the exact quantity). Prime Costs usually include Major Equipment, Bulk Materials, Engineering Hours and Construction Labour Hours. Derived costs are often factored costs (which as the name implies are based on applying a factor to the known cost elements) and also overheads and profit. Overheads and profit depend on the organisation and are not normally considered at the estimate stage. The sales and management teams will usually agree what element of profit to build in for a contracting situation and overhead recovery is generally set by annual budgets and applied evenly over all projects. A list of typical indirect cost is included in Appendix – 1.

1. Major Equipment

This will include Static equipment (Columns, Vessels, Tanks) Rotating Equipment (Compressors, Pumps) and their associated costs - Engineering, Procurement, construction (Erection, painting & insulation) and commissioning cost. Cost of internals, catalyst, first fill of oils etc. and vendor representative's presence during commissioning also need to be included.

2. Piping and associated cost

Engineering, Procurement, construction (Erection, painting, heat tracing & insulation), pre commissioning and commissioning cost.

3. Electrical

Cabling and cable trays, lighting, number of substations , power generators etc.

4. Instrumentation and Process Control

Engineering, Procurement, construction of instrument and process control related items e.g control valve, relief vale, cables, cable tray etc. including field auxiliary rooms, Control room and DCS related items.

7. Technical "know how" fees and licensed processes

Royalties for technologies have to be considered in the estimate.

8. Taxes and duties

Taxes and duties levied by governments in the location of the plant/project have to be given due importance in the cost estimate as sometimes they can increase the budget by up to 40%.

5. Civil and Structural

Engineering , procurement and construction of civil related items, including survey, land grading, paving, drainage, road, pipe rack, foundations etc.

6. Project Management Consultancy (PMC)

For a large project the owner will appoint a group of engineers to look after the project, the term for this is project management consultancy.

factor needs to be applied to the man-hours. The Productivity of labour in all geographical locations is not the same. Productivity is also affected by the skill of the workforce and the quality and quantity of supervision. It is advisable to consider that productivity will not be high to avoid budget over runs particularly as productivity inevitably reduces towards the end of a project. Productivity is expressed as a factor, 1.0 meaning that the labour force is going to perform 100% and then reductions are represented by less than 1.0. So if our productivity is estimated to be 20% less than perfect that is a factor of 0.8. It follows then that to complete 100 hours of work we divide by 0.8 giving an estimate of 125 hours to complete the scope.

9. Labour hours and costs

After the capital cost of equipment the most important item to get right is the quantity and cost of labour man-hours. Not just for the cost element but also as they will form the basis of working out the construction schedule for the project. Every construction activity has its own man hour component. The hours required to perform any activity can be found or calculated using one of the many readily available estimating norm books. Once we know the total number of man-hours these can be used to calculate the total workforce required. To do this calculation a productivity

10. Construction equipment and its insurance

Modern construction is mechanised and equipment intensive to reduce construction schedule and cost. There should be a good evaluation on the requirement (Buy/Hire) of speciality Cranes (Heavy Lift), Cranes, Excavators, Rock breaker, trailers, tools and tackle etc. The cost of insurance of all bought or hired equipment items also has to be considered.

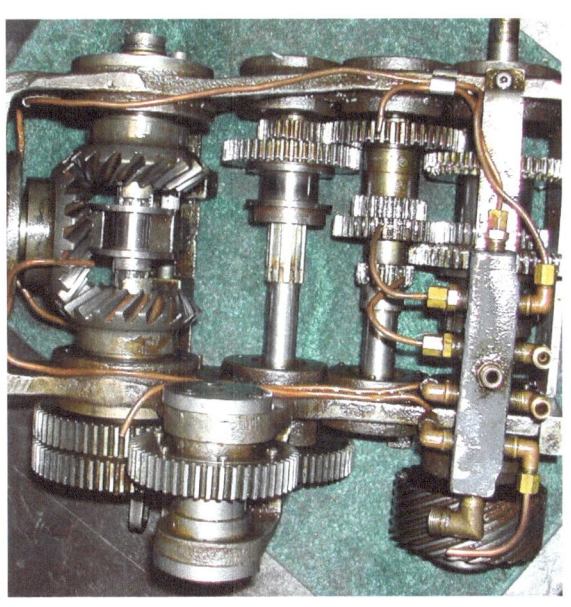

11. Temporary Construction facilities and indirect costs

For most projects before work commences at site it is necessary to establish essential services. These include fire water, drinking water (potable), First aid facility, access ways (roads) etc. A budget for the construction of and provision of these facilities needs to be included in the estimate.

12. Pre-commissioning and commissioning cost

There are several cost items in this group, e.g. hydro test water, Pneumatic testing/flushing blowing air compressor, chemical cleaning, steam blowing, helium leak testing, first fill process liquid which turns out to be waste etc.

13. Contingency

Once all the known facets of the estimate are catered for the unknowns have to be allowed. As is noted in the beginning chapter of this guide accuracy is one that needs to be allowed for. Then the variables to the estimate have to be identified and factored in. These variables will include elements such as claims by contractors, these may be a major or minor element dependant on factors such as imposed time constraints where additional work is added by the client due to process changes etc., likely interference from other contractors, failure of the client to provide services or utilities promised under the terms of the contract, failure to give access to the work site on time or not having permits in place for importation of goods into the country.

14. Insurance

As previously noted insurances of all kinds need to be considered. Construction All Risks insurance (CAR) to cover the damage to the plant caused by a major construction related event is required. The cost of this insurance will be quoted by insurance companies based on the estimated total cost of the project. This insurance may be placed by the client on major projects. The insurance will have a deductible which needs to be insured by the contractors or an allowance for several events up to the limit needs to be included in the contingency.

Modern construction is mechanised and equipment intensive to reduce construction schedule and cost. There should be a good evaluation on the requirement (Buy/Hire) of speciality Cranes (Heavy Lift), Cranes, Excavators, Rock breaker, trailers, tools and tackle etc.

Chapter **4**

Methods
of Cost Estimation

There are several ways to estimate each component of a cost estimate. The below methods are popular and widely used.

1. Costs from vendors
2. Costs from previous similar projects
3. Deriving costs from empirical formulas and the cost of raw materials

The above three methods are described below.

Cost from Vendors

The safest way to estimate a cost is to ask a vendor for a budget quote and use the vendor provided price as the estimated value.

There are advantages and disadvantages to this method.

Advantages

1. Dependable

70-80%. Budgetary quotations from manufacturers are reliable. As the manufacturers know the raw material price and are also aware of the trends of the raw material markets, their budgetary quote should therefore be fairly accurate.

2. Practical

The price provided by vendor will be far more practical and close to the purchase cost than any other estimated method.

Disadvantages

1. Overhead linked

The price quoted may vary by 20-30% from one vendor to the other. It depends on several factors. The overhead of the manufacturing company is one of the significant factors to control the finished product price. The more the overhead, the higher will be the quoted price.

2. Vendor Capacity

If the vendor's shop is heavily loaded, the vendor may give a higher budget quote, often to cover his own costs in rescheduling his manufacturing output to accommodate your delivery requirement or simply because he sees it as an opportunity to increase his profit margin.

3. Location

An item manufactured in third world countries will be cheaper than the same item manufactured in Europe for instance. This is because of the relative costs of labour. Of course quality needs to be assured. The Estimator needs to take this in to account.

Cost from a previous similar project

Whether for a client Company or an EPC Contractor, real project costs from a recently constructed similar project is a good source of cost data. The costs from these sources are "real world" and include for recent market conditions. They will also give a good indicator for the escalation of costs from the Control Estimate to the Final Forecast. Of course how the execution contractor performed and what curve balls were thrown during the course of the project will all have an effect. The costs will therefore need to be normalised to fit to varying project locations for example and also to allow for execution strategy and foreseen changes in global markets.

Deriving cost from empirical formulas and cost of raw materials

Sometimes estimators do the basic design of an item using engineering design formulas. After obtaining the basic weight, volume etc., estimators use standard multiplication factors published in cost journals to obtain the estimated price.

This method is good for comparison purposes or where there is no readily available cost data. However the first two methods are more dependable and are widely used.

After establishing the components of a cost estimate an estimator has to assign a price to each component. There are many methods to finalise the price for each component. There is no fixed rule for this. This is up to estimator to follow a method. If the project being estimated is to be built locally it is a common practice to take the quotations of local vendors (prospective project approved vendor). If the project is local and to be built in future the local commodity inflation trend may be applied to the cost estimate. In case the project is to be built in a different country altogether prices are normally obtained either from past records of that country or an agreed conversion factor to other locations. Collection of cost data from a cost bank is very important. The bigger the cost data bank is the lesser the chance of a cost overrun. It is to be ensured that the relevant tax & duties, regulations, insurance conditions of the new location are followed.

The safest way to estimate a cost is to ask a vendor for a budget quote and use the vendor provided price as the estimated value.

Chapter 5

Project Risk

and Cost Estimation

In a very simple manner, project risks can be categorised as below.

1. Project Scope
2. Regulatory rules
3. Market
4. Customer
5. Subcontract & Suppliers
6. Budget
7. Schedule
8. Funding
9. Resources
10. Contracting Strategy
11. Industrial Relation/Health & Safety

To a cost estimator, every risk is additional cost. Any EPC contractor has to mitigate the risk in the project cost.

Project Scope

The risk of project scope may include but not be limited to, a change in project specification, change in technology and change in quality standards. The cost component may be 1 to 2% of the total project cost

Regulatory Rules

For most projects it is very unlikely that regulatory rules will ever change during a project life cycle. Larger longer term projects can be affected. An example would be a local authority putting a higher specification requirement on the building of labour camp. Cost components may be 2-3% of total project cost.

Market & Customer

Change of market and customer are very rare. If it happens the marketing strategy would need to be revised. Cost component is typically 0.5% of total project cost.

Subcontractors and Suppliers Budget Quote Rise

While estimating a project cost, it is common practise to obtain the price of commodities or services from Subcontractors and suppliers. It is expected that these Subcontractors and suppliers will be available later on for actual work. The risk is that they may not be available or increase their bid price. Cost component is 0.5% of total project cost.

Budget and Schedule Overrun

This eventually happens to most of the mega scale projects. Cost components may be 5-10% of total project cost, depending on project size and criticality. These are also affected by contract strategy, refer below.

Funding

How you fund your project, which currency you work in and how you hedge those currencies in the money market will have an effect on the overall project cost. A risk assessment needs to be carried out appropriately. Cost component is likely to be 2-3% of total project cost.

Resources

Normally the availability of resources does not have a large impact on a project. Labour markets are generally flexible enough in most industrialised areas to ensure there are sufficient supplies of skilled workers. However there have been years where due to the number of projects worldwide the number of available skilled resources worldwide was not enough. These conditions lead to paying a premium on rates and possibly having to teach/train up new skilled workers. Cost component is about 0.5% of the project cost.

Contracting Strategy

Several types of contract exist. The two main types occurring most regularly are Lump Sum and reimbursable/ unit rate.

Lump Sum is often seen as the best option for ensuring the costs are fixed as in theory you pay a set price for the work you want and you are more likely to hit your schedule target as delay is at the contractors risk. Reimbursable or unit rate contracts increase or reduce in cost according the scope changes and these are often seen as too volatile.

Actually Lump Sum is often a riskier option. If the design is not complete, changes will be expensive. Also there is less flexibility, Lump Sum contracts will have a schedule agreed which if disturbed by for instance , activities of other contractors on site or a failure of the client to provide promised services or utilities on time it will lead to extension of time claim and cost.

Reimbursable or unit rate contracts have the advantage of flexibility. Changes of scope can be estimated and agreed using set rates and interference from the client or other contractors rarely results in traditional claims. The price and durations are more easily adjusted to suit. The big disadvantage however is that the schedule is always at risk. There is little incentive for the contractor to ensure that the end date is met. These types of contract require more managing by the client and therefore by nature require the client to invest more cost in a bigger project team.

Industrial Relation/Health & Safety

Good industrial relations and a robust health and safety principle are vital to any project. Without this, labour unrest, lost time due to injuries and bad working conditions can lead to labour disputes and low morale in the workforce, resulting in low productivity or at worse vandalism and sabotage .This risk should be managed carefully and including enough cost for adequate manpower and facilities including safety incentives will mitigate this risk.

Chapter **6**

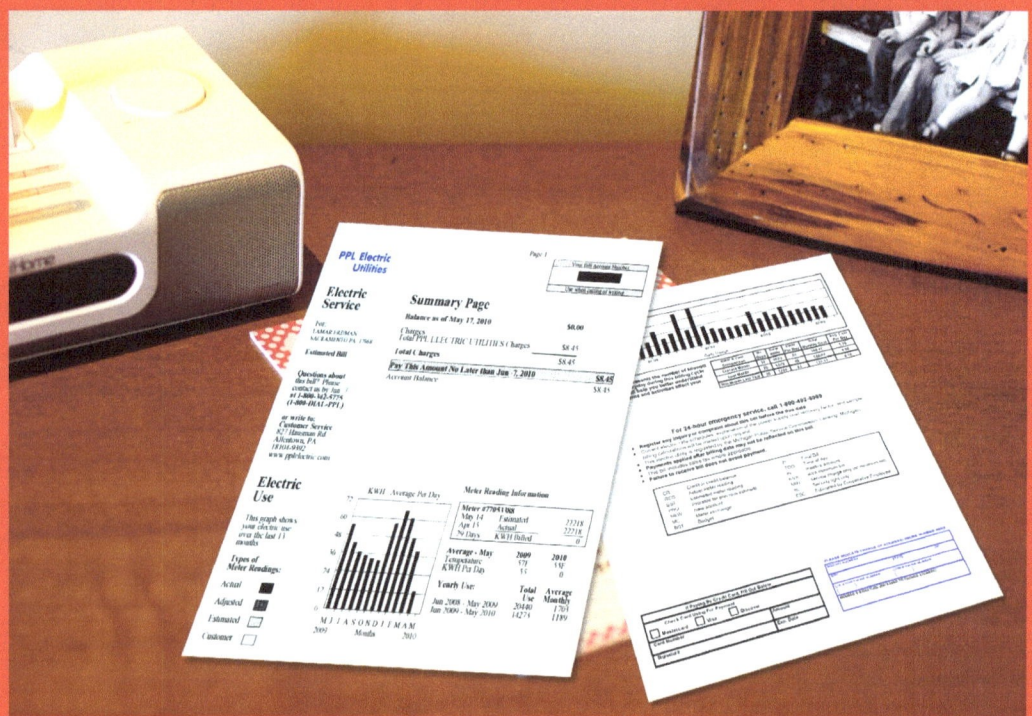

Examples
of Cost Estimation

ITEM	Quantity	UNIT COST US $	Total Estimated Cost US $
PIPING ESTIMATED COST FOR THE FOLLOWING			
P-123 A/B NEW BIGGER SIZE PUMP	2	12391	24781
P-456 C ADDITIONAL PUMP	1	5489	5489
P-789 C ADDITIONAL PUMP	1	4535	4535
P-147 A/B PUMP IMPELLER / BASE FRAME CHANGE	2	1000	2000
P-148 A/B PUMP IMPELLER / BASE FRAME CHANGE	2	8728	17455
P-245 IMPELLER BASE FRAME CHANGES	1	2000	2000
E-234 ADDITIONAL KERO EXCHANGER	1	175142	175142
Material cost of steam tracing, painting and insulation = 5% of total piping Material Cost			11570
SUBTOTAL PIPING + INSULATION & PAINTING			**242972**
INSTRUMENTATION			
123 XV 12/34 (10" ALLOY STEEL VALVE)	2	10000	20000
134 PV 45 / 56 (12" CS CV)	2	9000	18000
456 TV - 345 (4" CS CV)	1	2000	2000
Instrument Bulk Material cost = 2.5% of Tagged item cost			1000
SUBTOTAL INSTRUMENTS			**41000**
SUBTOTAL INSTRUMENT + PIPING ====>			**283972**
Additional 25% TOWARDS TAX, DUTIES AND TRANSPORTATION etc.			70993
MATERIAL COST			**354965**
EXPEDITING/INSPECTION COST (5% OF TOTAL MATERIAL COST)			17748
MATERIAL + INSPECTION COST			**372713**
CIVIL			
CIVIL SUPPORTS = 5% OF PIPING COST			12149
ADDITIONAL FOUNDATION ETC IN M3	50	150	7500
ENGINEERING COST			**19618**
(5% OF TOTAL COST)			
MATERIAL + INSPECTION + ENGINEERING COST			**411980**
EXECUTION			123594
30% OD TOTAL MATERIAL COST			
GRAND TOTAL			535574
TOTAL COST		**SAY**	**5,40,000 US $**

Exclusions: 1. Pump + Impeller + Motor

Above is one practical example of cost estimation of an expansion project (EPC) together with estimations thumb rules.

During this expansion project the below items are intended to add.

1. Piping for Pumps and motors – 06 no's Tagged items (free issue)

2. Heat exchanger – 1 no

3. Instrument – 3 tagged items

4. All other support items to engineer, procure and installation of the above items.

The assumptions used in the above estimation are as below:

- In many projects, we notice that insulation and painting are not properly estimated and as these items are very much labour intensive job, it eats away the profit percentage. In this above estimation, the assumption of material cost of steam tracing, painting and insulation = 5% of total piping Material Cost.

- Tax, Duties and Transportation cost = 25% of the total materials cost

- Expediting/Inspection cost = 5% of total material cost

- Engineering cost = 5% of total (material + Inspection + tax & Duties + logistics) cost

- Construction and contracting cost = 30% of total (Material + Engineering) cost

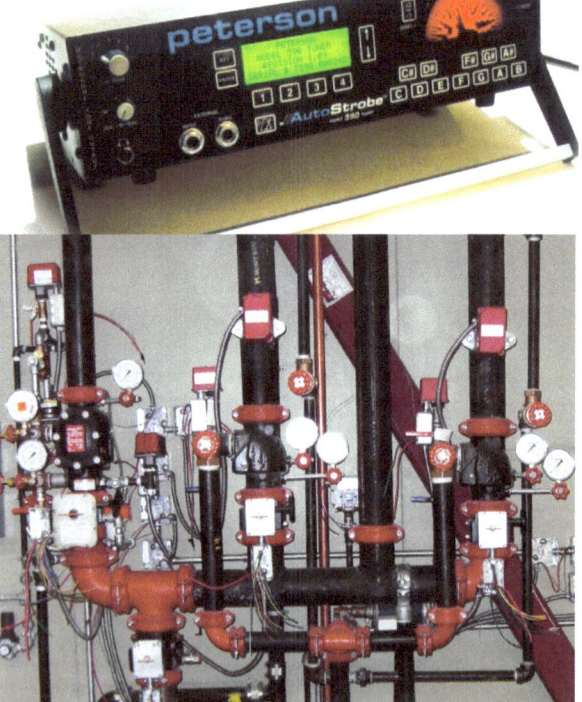

Chapter **7**

How to make Estimation
'Rules of Thumb' from practical examples

Y ou might have heard and read many thumb rules on cost estimation . There are many rules of thumb available in the industry , which can be difficult to verify . It is tough to trust any of the thumb rules unless you know the procedure used to produce it. If you have experience and gathered data for a particular industry, you can make thumb rule by yourself and apply it very well. Here below is a simple example, how a thumb rule can be made.

This example has been derived from our past experience of engineering design of EPC projects of process plants.

First column is the various departments of an EPC company, against which man-hours are generally booked. Second column is the actual man-hours booked and the third, the last column is for % booking of man-hours against total hours.

In an EPC company supporting department's man-hours cannot usually be booked directly to the client. However the support services are necessary to run the project. The trick is to spread all support services hours among engineering disciplines. This will help accurately budget the project and minimise risk of cost overrun.

The abbreviations used below tables are as under.

PR	=	Process Engineering
PI	=	Piping Discipline
INST	=	Instrument Engineering Discipline
EI	=	Electrical Engineering Discipline
ME	=	Mechanical engineering Discipline
CI	=	Civil Engineering Discipline
PROJ	=	Project Engineering/Project Management Discipline
PL	=	Planning Discipline
IT	=	IT Infrastructure Services
DATA	=	Electronic Storage and Data Management.
QC	=	Quality Control
DOCU	=	Document Management
SECRE	=	Secretarial Support

Examples of 'Rules of Thumb'

1. Engineering is 10%-15% of Total Installed Cost (TIC).

2. Split of Procurement (Materials Equipment) to Construction is 60/40 to 70/30.

So a typical project split could be:

15% – Engineering

51% – Material Procurement

34% – Construction/Commissioning

Step – 1

Step 1 is to put actual values against all engineering disciplines and support services as well.

DISCIPLINE	MANHOURS	% ACTUAL		
PR	1800	9.8%		
PI	4550	24.7%		
INST	1375	7.5%		
EL	1800	9.8%		
ME	1684	9.1%		
CI	4300	23.3%		
PROJ	750	4.1%		
PL	450	2.4%		
IT	320	1.7%	LIQUIDATING INTO 6 DISCIPLINES	
DATA	650	3.5%		
QC	100	0.5%	1720	
DOCU	450	2.4%		287
SECRE	200	1.1%		
	18429			

Step – 2

Step 2 is to transfer the loading of Project and Planning services to engineering disciplines, based on % of loading.

DISCIPLINE	MANHOURS	% MODIFIED		
PR	2087	11.3%		
PI	4837	26.2%		
INST	1662	9.0%		
EL	2087	11.3%		
ME	1971	10.7%		
CI	4587	24.9%		
	750	4.1%	LIQUIDATING INTO 6 DISCIPLINES	
	450	2.4%		
			1200	
				200
	18429	100.0%		

Step – 3

Transfer the balance % loading of support services to engineering disciplines.

DISCIPLINE	MODIFIED MANHOURS	% MODIFIED
PR	2287	12.4%
PI	5037	27.3%
INST	1862	10.1%
EL	2287	12.4%
ME	2171	11.8%
CI	4787	26.0%
	18429	100.0%

Now, the above table is ready for use for estimation of man-hours.

Chapter **8**

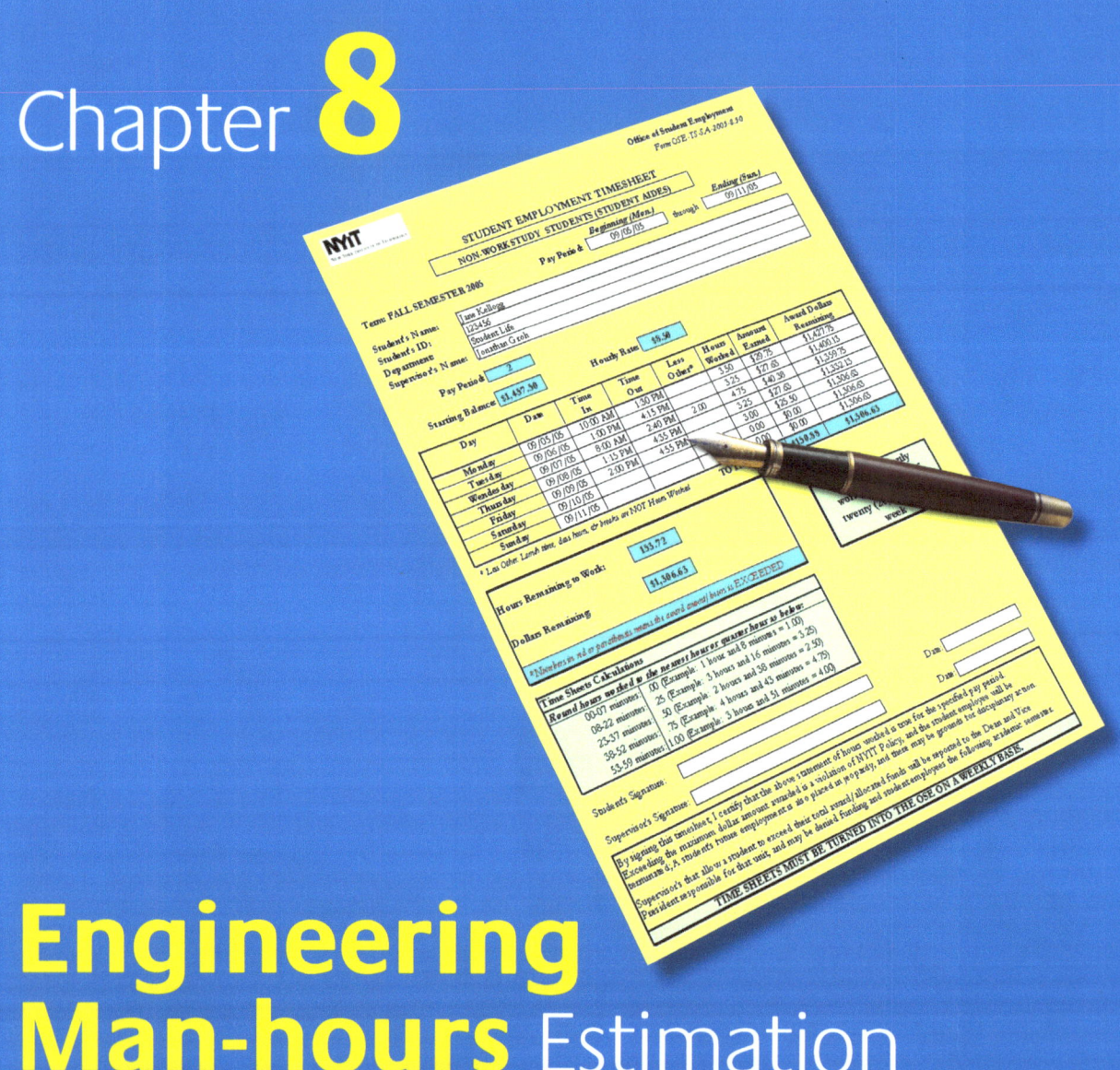

Engineering Man-hours Estimation

'Rules of Thumb' example

Discipline wise % man hour distribution of a typical project is described below:

Discipline	% Man-hours
Process Engineering	10
Static Equipment	
Rotating Equipment	15
Package Equipment	
Electrical Engineering	10
Instrumentation Engineering	15
Piping layout	
Piping Design	30
Piping Stress and Support	
Civil & Structural Engineering	20

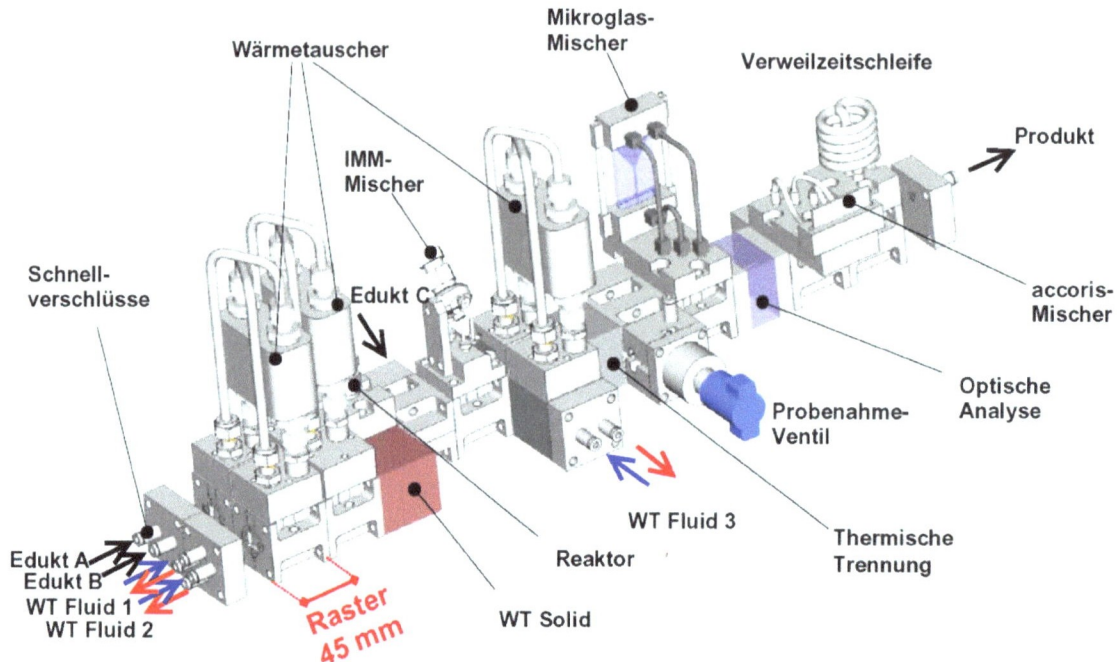

Typical Man-hours of Standard Process Engineering Deliverables are described as under:

Process Deliverables	Man–hours/unit deliverable
PFD	120
P&ID	60
Line List	100
Equipment List	60
Process Datasheet of all Equipments	20
Control valve, Flow element, Safety Valve Process datasheet	12
Process Description/Operation Manual	400
Hazop and any other safety review	80

Typical Man-hours of Standard Piping Engineering Deliverables are described as under:

Deliverables	Man–hours/unit deliverable
Plot Plan	120
Equipment Location Plan	60
Line List	100
3D model (Per P&ID) including clash check	150
Piping general arrangement drawing	30
Isometrics	15
Flexibility Analysis and support (Per Line)	40
Inspection and test plan	20
Material Take off preliminary, intermediate and final take off	80
Requisitions – preparation, review, vendor co-ordination, Technical Bid analysis and approval of vendor drawing	80

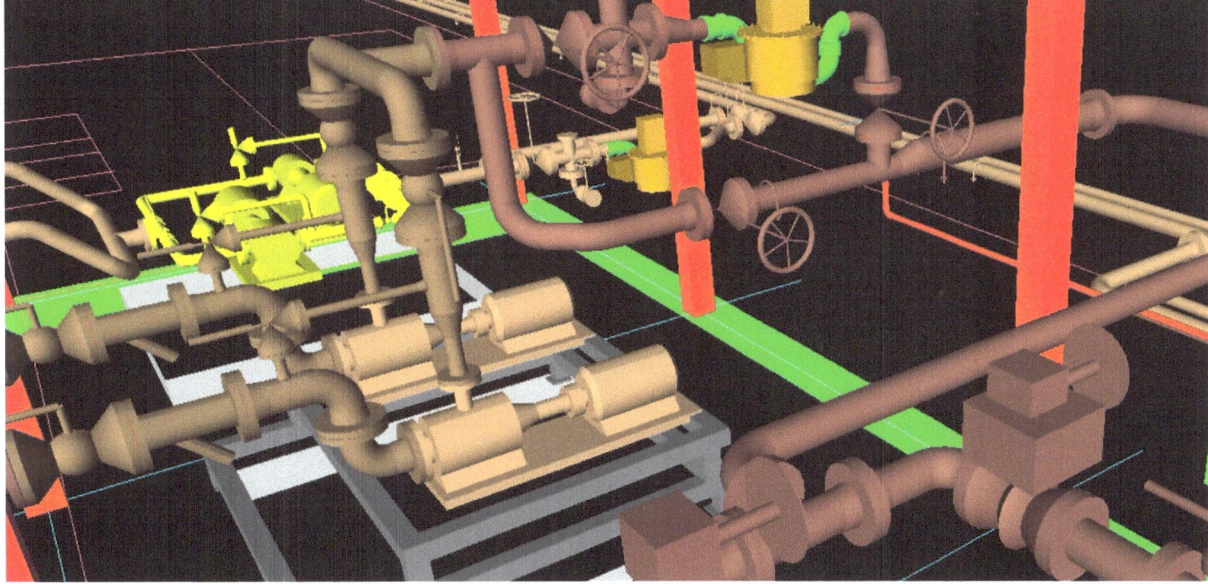

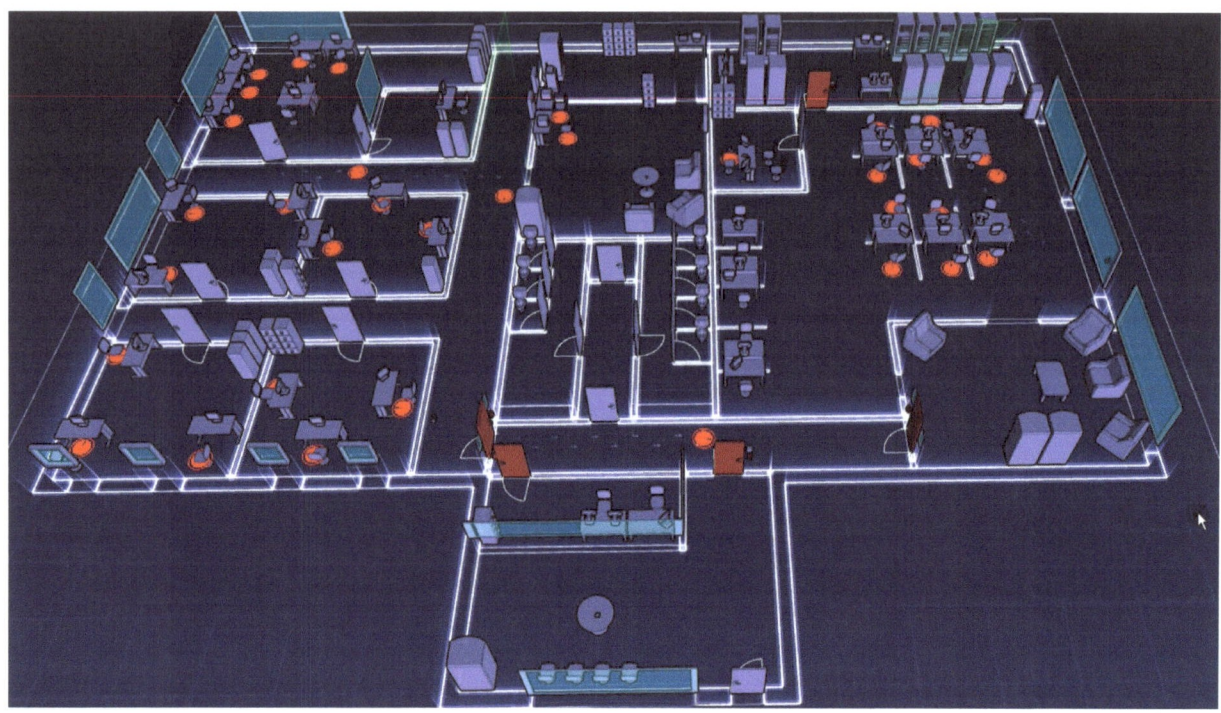

Typical Man-hours of Standard Mechanical Engineering Deliverables are described as under:

Deliverables	Man–hours/unit deliverable
Datasheet	40
Inspection and test plan	20
General arrangement drawing	40
Nozzle orientation drawing	20
Vessel body Weld location drawing	40
Platform and piping clit location drawing	40
Requisitions – preparation, review, vendor co-ordination, Technical Bid analysis and approval of vendor drawing	80

Typical Man-hours of Standard Electrical Engineering Deliverables are described as under:

Deliverables	Man-hours/unit deliverable
Datasheet	20
Inspection and test plan	20
Single line diagram	40
Power layout	40
Power distribution schedule	20
Motor list	40
Hazardous area classification drawing	40
Cable specification	40
Cable sizing and selection study	40
Cable schedule	40
Substation layout	40
Cable layout	40
Cable tray layout	40
Cathodic protection layout	20
Earthing layout	40
Lighting Layout	40
Telecommunication layout	40
Electrical standard installation details	20
Distribution board , protection panel schedule	40
Electrical equipment Layout	40
Termination and wiring diagram	30
Motor starting study	20
Factory acceptance test procedure	20
Painting, packing and shipping Procedure	20

Typical man-hours of standard Instrumentation and Process Control deliverables are described as under:

Deliverables	Man–hours/unit deliverable
Instrumentation Index	40
Inspection and test plan	20
Instrumentation Datasheet	20
Process Control Narrative	50
Cause & effect diagram	40
Cable layout	40
Cable schedule	40
Cable specification	20
Bill of material	160
Requisitions	20
Inspection and test plan	20
Cable reel handling procedure	20
Cable tray layout	40
Instrument Plot Key plan	40
Instrument hook-up details	20
Impulse line hook-up details	20
Instrument layout key plan	40
Junction Box schedule	40
Junction box location diagram	40
Termination and instrument Wiring layout	40
Field auxiliary room and analyser house instrument layout	40
Patch panel and system cabinet layout.	40
Cabinet wiring schedule	40
Fire and gas system layout	30

Deliverables	Man–hours/unit deliverable
Public announcing system layout	30
Alarm and trip schedule	40
Loop drawing	10
Intool Database	160
Process graphics, F&G graphics.	15
Fire & Gas - logic details design specification	40
DCS – Basic control details design specification	40
DCS sequence , special control FAT procedure	20
DCS – Human machine interface FAT procedure.	20
System and marshalling hardware FAT procedure.	20
Alarm Management procedure	20
Instrument earthing philosophy	20

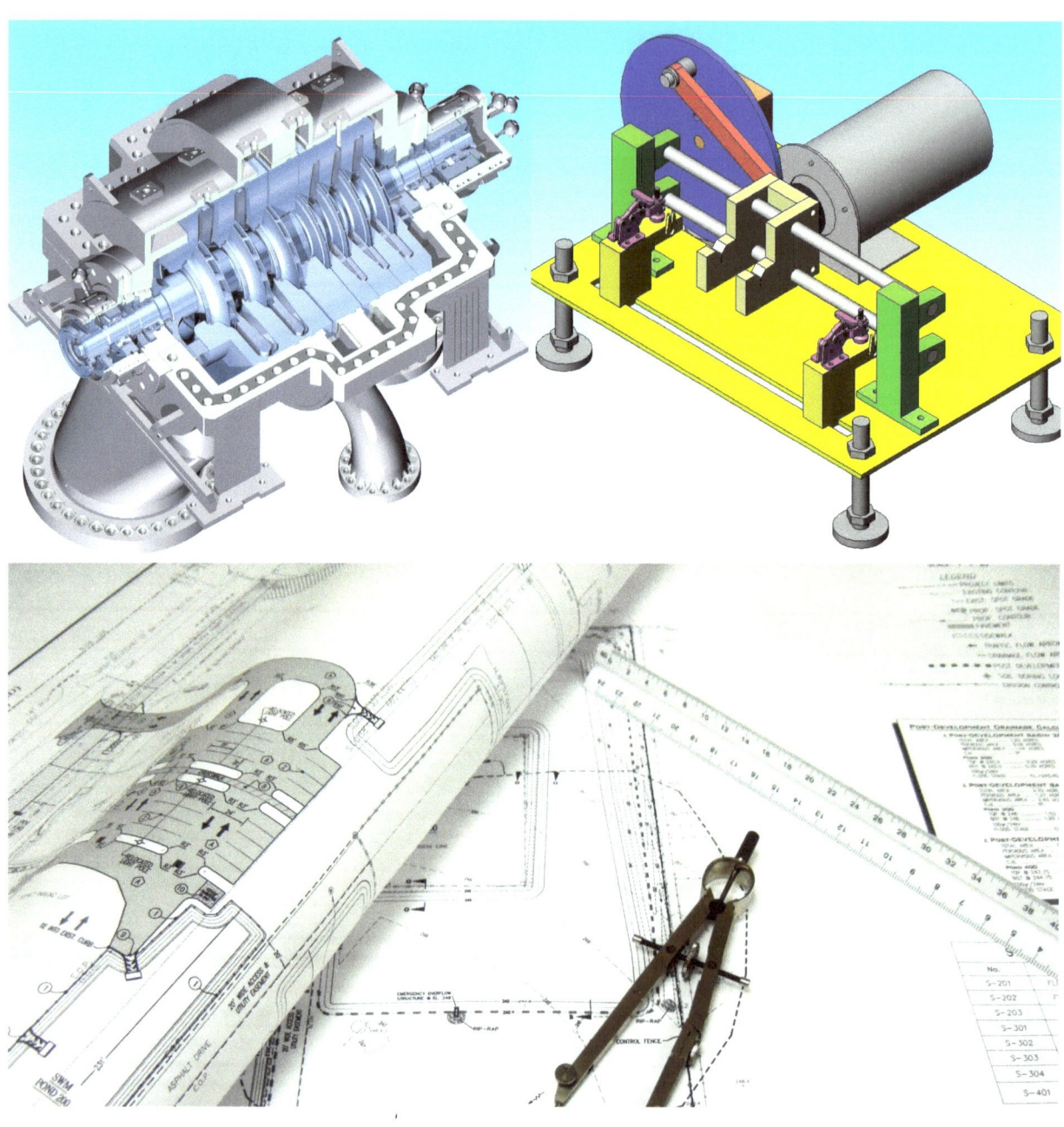

Typical man-hours of standard Civil Deliverables are described as under:

Deliverables	Man-hours/unit deliverable
Foundation Drawing	20
Inspection and test plan	20
Foundation layout	40
Structural framing plan and elevation	25
Structural foundation plan	
Under ground drainage system detail drawing and layout	60
Underground general arrangement drawing	40
Building related drawing	30
Substation plan elevation and section	35
Field auxiliary room plan elevation and section	35
Paving key plan and detail paving layout	40
Embedded plate layout for earthing connection	25
Beam pin connection drawing	25
Bracing , ladder , handrail and grating floor detail drawing	20
Pipe rack frame and elevation drawing	25
Equipment structure frame and foundation plan	25
Pipe rack embedded plate details	25
Shelter details	16
Pipe rack floor opening details	16

Chapter 9

References

**Basics of Multi-Discipline Project Engineering
by S. Moulik, 2010**

Appendix 1

List of typical indirect costs to be considered in an estimate

Travel	Mock-Ups
Temporary Construction Site Mobilisation	Testing & Inspection
Construction Photographs	Soils/Geotechnical Testing
Safety Training & Orientation Equipment & Materials	Environmental Testing
Safety Consumables	Temporary Construction Power
Environmental Consumables	Electric Bills
First Aid/Medical Services	Temporary Fire Protection
Safety Programs & Initiatives	Temporary Gas/Oil Heating Bills
Fire Protection	Temporary Heaters
Safety Signs	Telephone Bills
Craft Appreciation Events & Awards	Telephone/Data Lines
Office Supplies	Water Bills
Office Equipment & Furniture	Temporary Water Lines
Computer Hardware	Sewer Bills
Computer Software	Temporary Sewer Lines
Software Training	Temporary Lighting
Computer Support	Field Office Construction
Postage & Messenger Service	Sanitary Facilities/Toilets
Plans, Prints & Reproduction	Trailer Rental
Project Team Expense	Lunch/Break Facilities
Licences/Permits	Temporary Structures – Other

Job Services	Temporary Heating Enclosures
Temporary Elevators (Provisional Sum)	Temporary Weather Protection
Temporary Hoists	Temporary Surface Water/Runoff Controls
Temporary Cranes	Dust Control
Temporary Radio	Special Area of Conservation (Sac) Preservation Controls
Pickup Trucks	Temporary Directional/Construction Signs
Lifts (Fork Lifts/Scissor Lifts)	Mob/Demob Equipment
Equipment Rental – Other	Erect/Dismantle Equipment
Transport Equipment	Fuel, Oil, & Maintenance
Access Roads	Warehousing
Temporary Roads	Field Survey
Temporary Parking	Topographical/Boundary Survey
Work Bases/Lay down Areas	Site Maintenance
Pedestrian Pathways	General Clean Up
Temporary Barriers	Skips
Security Shacks/Gates	Recycling Skips
Security Patrol/Guards	Office Maintenance & Cleaning
Security Equipment & Supplies	Final Clean-Up
Temporary Protection	Fees & Insurances

Appendix 2

ESTIMATE FLOW DIAGRAM – 1

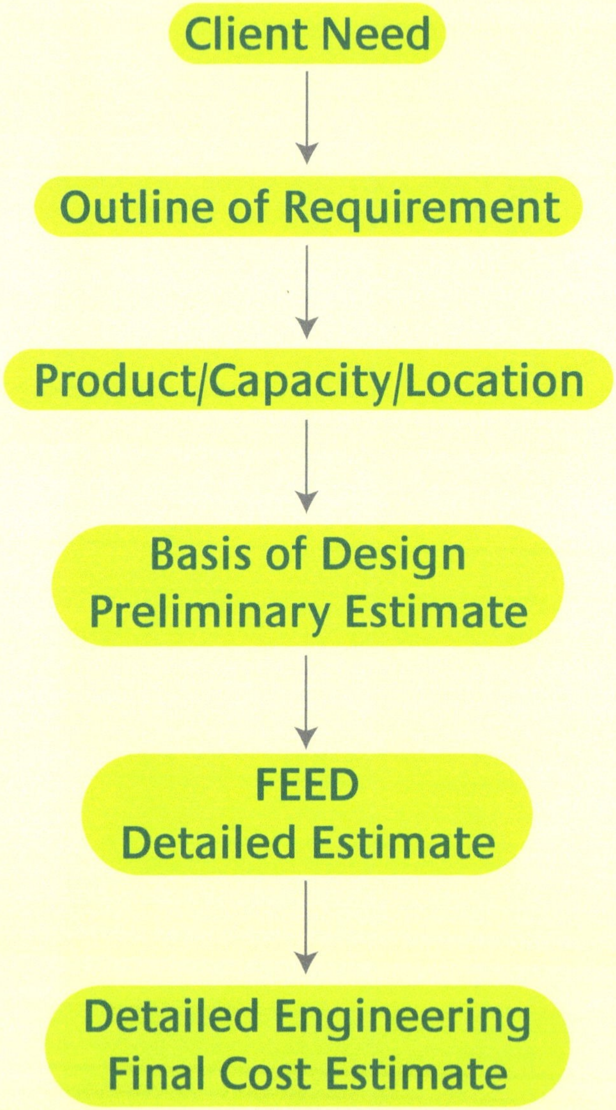

Client Need

↓

Outline of Requirement

↓

Product/Capacity/Location

↓

Basis of Design
Preliminary Estimate

↓

FEED
Detailed Estimate

↓

Detailed Engineering
Final Cost Estimate

ESTIMATE FLOW DIAGRAM – 2

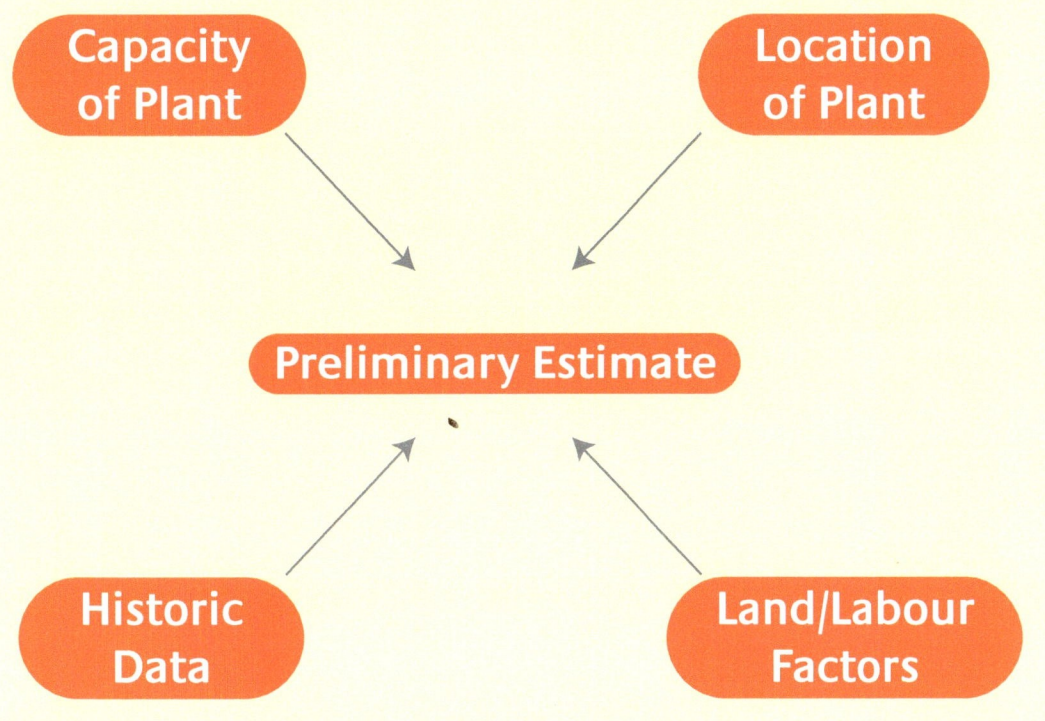

ESTIMATE FLOW DIAGRAM – 3

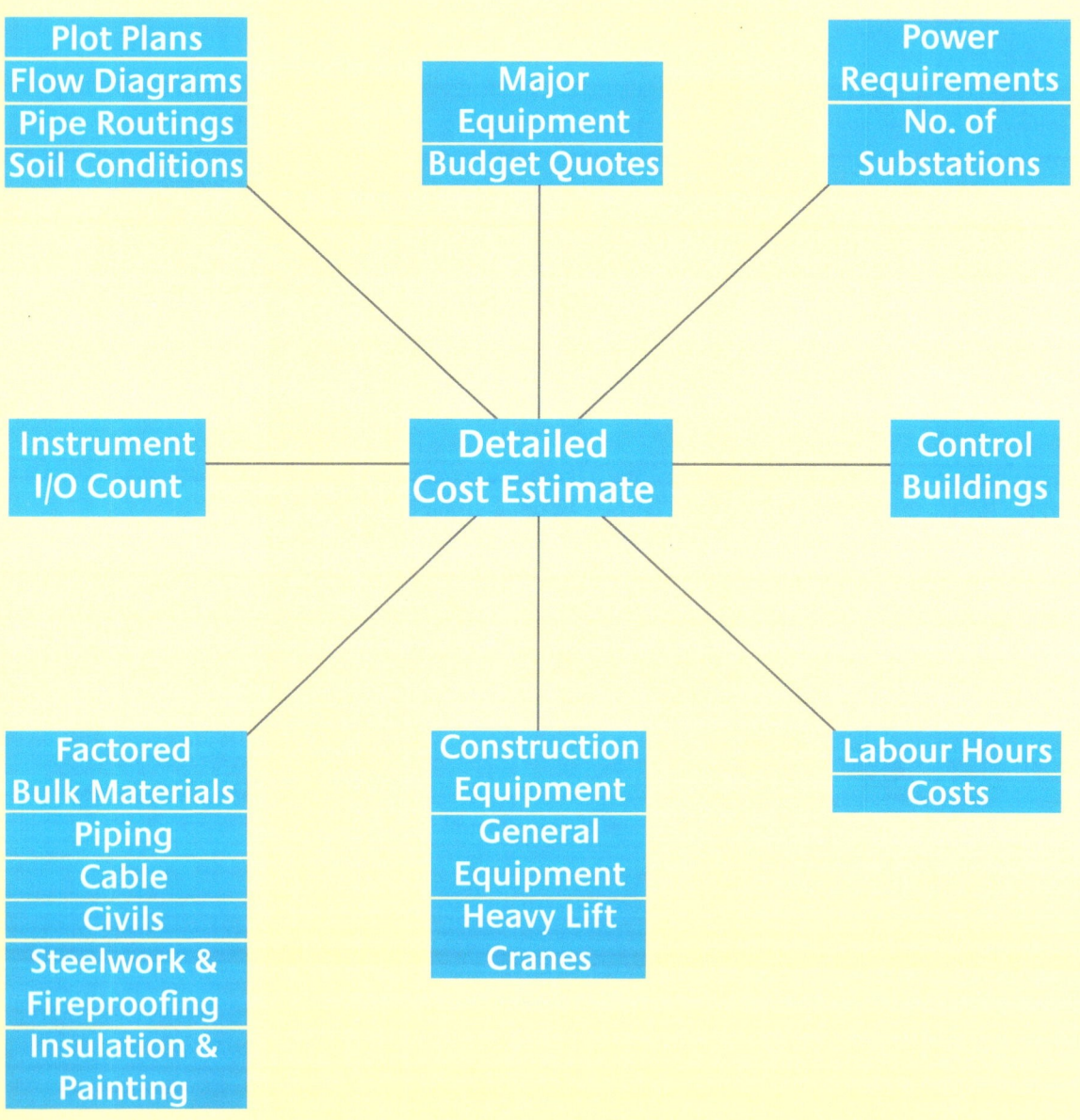

ESTIMATE FLOW DIAGRAM – 4

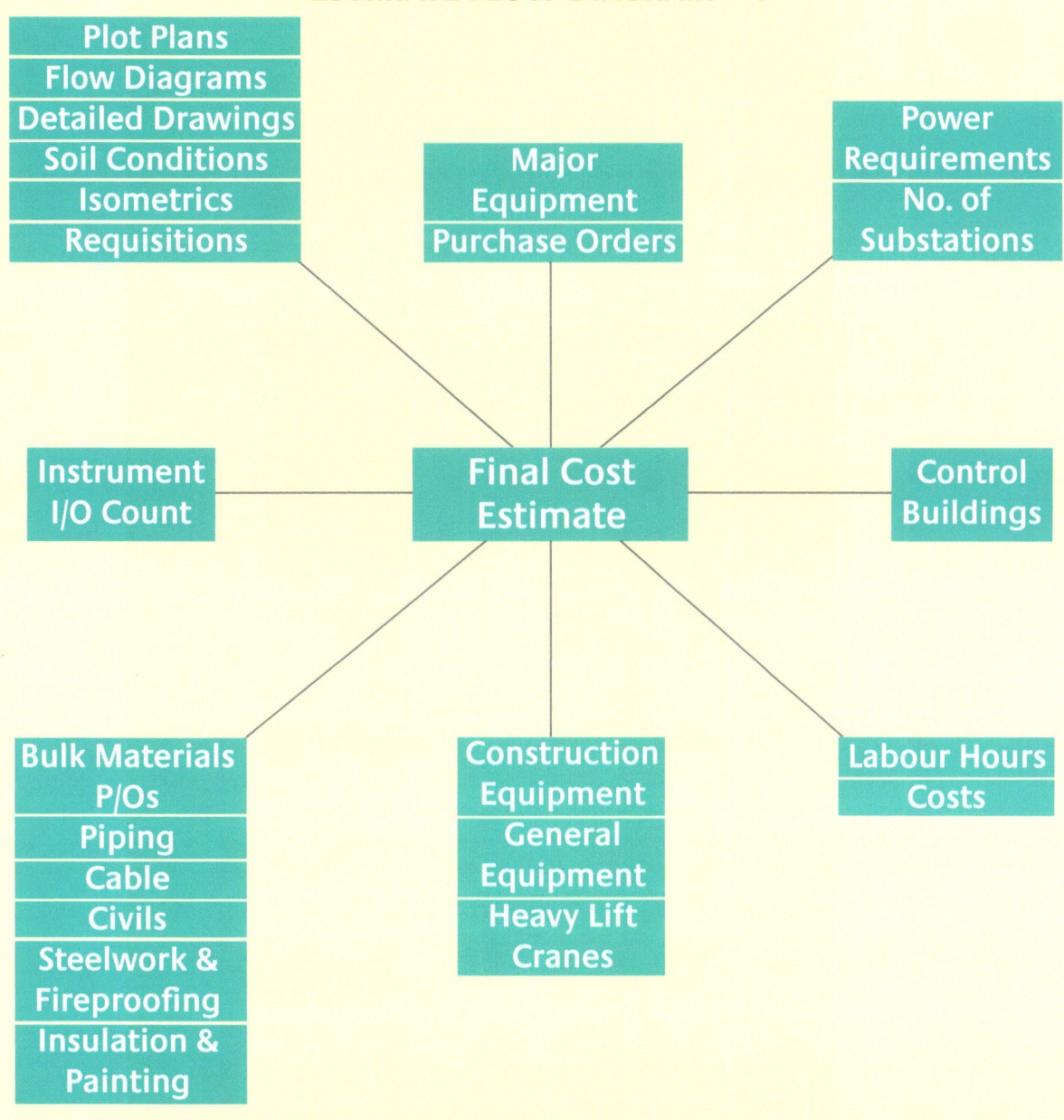